the Catcher in the Sky

Astronomy Story of the CLOSER KIND

Christopher J. Rosin

TABLE OF CONTENTS

INTRODUCTION...3

Part 1: HEAVENLY DISCOVERY.......................................5

Part 2: FAINT FUZZIES 9

Part 3: STAR WARS 14

Part 4: SPACE AND TIMEWARP.....................................20

Part 5: VISUAL AWAKENING..25

Part 6: FRIENDS INDEED...29

Part 7: OBSESSION 37

Part 8: MISSION TO MARS 40

Part 9: BIG GUNS 46

Part 10: CAPTURED PHOTONS..48

Part 11: SOLAR TOTALITY 58

APPENDIX...64

INTRODUCTION

Congratulations and thank you for your purchase of the paperback copy of my book. It was my pleasure to write this and hopefully provide you with a good blend of my astronomical experiences, purchases and challenges. I hope to also convey the extraordinary convergence with the possibility of our extraterrestrial friends along the cosmic journey.

I dedicate this book to the love of my life, my wife. She's the sweetest heart, my muse, and my wonder-woman. She's lived through my unremarkable past and makes my present the best gift ever.

Inspiring Look-backs
Life-changing events: The Moon Landing created the little astronaut in me. The Hubble Telescope, Space Shuttle launches, technology, and accidents showed that even rocket science is fallible. I can't wait for the next space telescope and the next manned missions to the moon.

Books that inspired me to write: "The Kite Runner" by Khaled Hosseini, "Stranger in a Strange Land" by Robert A. Heinlein, and, of course, "The Catcher in the Rye" by J. D. Salinger.

Songs that struck a chord when I was growing up: "Calling Occupants of Interplanetary Craft" by The Carpenters, "Children of the Sun" by Billy Thorpe, "New World Man" by Rush, and "The Great Gig in the Sky" by Pink Floyd.

TV Shows that made my imagination run wild: "The Twilight Zone," "Lost in Space," "The X-Files," "Cosmos," "Battlestar Galactica," and "Star Trek."

Movies that hit a home run: "Star Wars," "2001: A Space Odyssey," "Contact," "Starman," "Interstellar," "The Thing" (1982), "Ad Astra," "War of the Worlds," original release, "Close Encounters of the Third Kind" and "Greenland." Documentary: "The Billy Meier Story."

Part 1: HEAVENLY DISCOVERY

Let the stellar game of life begin...

It was a cold winter night. The kind of night where you see your breath appear before your eyes, and you crave a nice cup of hot chocolate with marshmallows. Or as the British say, "It's cold to your bones, sir."

My grandad said, "Let's look through your telescope, remember the one Santa just brought you? Let's see what we can see. Whatcha say, old boy?" He was an Aerospace Engineer with a secret clearance and loved science. I said, "Okay, let's, grandad, this is going to be so keen and out of sight." I was 9 years old and already had little stars in my eyes.

He assembled the scope, put in an eyepiece, and locked onto the brightest blue star in the sky (Sirius). He focused it and asked me if I wanted to take a look? I said, "Heck yeah, I do, grandad."

I saw a bright but small blue light. It resembled a blue pinhole on my Lite-Brite toy. I thought to myself, "Why does my grandad like this little blue dot? It's a pinhole, for goodness' sakes." I was discontented with such a tiny star. My grandad explained, "Here, I have old eyes, old boy; you can focus it to your eyes using the focuser." He then showed me how to rotate the focuser in and out of focus. I said, with excited amusement, "So keen, grandad, now that's a big beautiful, bright blue star. It reminds me of Christmas. Has it always been there? Was it there with Jesus?" Even though my impression of the star was literally out of focus, my brain was now focused and filled with so many questions. He was happy with my enthusiasm; however, when he looked through the eyepiece, he exclaimed, "Now our eyes are very different, you made it a big blue fuzzy ball for me, shshsh."

He laughed. I said, "I can see more when it's big like that, grandad." We then both laughed.

That little department store telescope was the beautiful beginning of my passion for the great heavenly sky-show of the unknown. I remember thinking, "I need a bigger telescope and all, but mostly, I want to catch a falling star, and then all my dreams will come true..." As you can see, at a very young age, I was already a little dreamer.

That same year, I had what I think was a close encounter of the third kind. My younger brother and I approached each other some ten years later about a dream we had of a metallic craft that stopped us in our tracks during an afternoon playtime with some neighborhood kids. We all looked up, pointed at it, just outside of our North Hollywood house, as it just hovered there. More on this extraordinary experience later in the Appendix.

Fast forward, five years later. I'm now 14 and a half years old. My grandma had taken me to go see the King Tutankhamen exhibit at the Natural History Museum. Another one of my favorite things to do, geek out at the museums. My favorite one to visit was the Science and Industry Museum.

My grandma started talking with some total strangers, like she had known them for years. She had a knack to strike up a conversation like this, with just about anyone, the gift of gab. She used to be an entertainer of sorts, so it was a no-brainer and deep-rooted in her, you know. She was a very colorful soul and had an infectious vibe about her. Everybody wanted to be around her. Holidays were a blast, as her sister's family would meet with ours, and the sisters would put on costumes and do these entertaining acts. I guess I have a small piece of that personality in my heart as well, or at least I'd like to think I do. She also was into studying the paranormal a bit. This is all a story for another time.

The woman she was talking to was with her young adult son, Jim. It turns out, Jim was an artist. His passion was acting. He had acted as a child and co-stared in, "Illustrated Man." He started talking to me and asked, "What do you want to be when you grow up?" I said, "An Astronaut, of course." He laughed and said, "That's what I wanted to be when I was asked that same question. When I was your age, I was discovered, and this acting thing kept getting in the way." Then he quoted from a movie, "*When I was your age, there were only the great diamonds and sapphires and emerald mists and velvet inks of space, with God's voice mingling among the crystal fires.*" He said, "Do you know what movie that's from?" I said, "No, but I think you're funny." We laughed and talked, and I was amused that this grown-up found, well found that he could talk to me. I felt older, and I loved it and him. I mean, I felt something more than just a kid, and he listened to me. He did not always try to correct me or act like I didn't know anything.

Some days later, he started living at my grandparents' house with us. My mom and dad had separated, and we, my mom, brother, and I were now living in what I called the mansion in Van Nuys. I'm not sure of the arrangements that were made for my new friend's stay, but I thought it was so nice to have a big brother of sorts. My little brother and I were in awe of his impersonations and his piano playing.

We thought, "Wow, we have a movie star living here with us under the same roof, so cool." It was a little odd having a grown-up around, at first, especially during playtime. Honestly, we thought nothing of it after a while and got used to him being there as just our newfound friend or, this big goofy kid.

7

One day we started talking about LA's landmarks, and he asked, "Have you ever been to Griffith Park?" I said, "Of course, Travel Town and the Zoo are great!" He said, "Not that, silly rabbit, the boss Observatory up there at the top of Griffith Park's hill." I said, "I think so, maybe I have, but when I was a little kid. I don't remember it really and all." He said, "Little kid, what last week?" I said, "No, stupid, is it like Busch Gardens, like, can we totally ride on the Bumper Cars and Merry-go-round?" He said, "No, but they have a really big telescope to look through and a planetarium. They've also filmed many movies up there like 'Rebel Without A Cause.' Wanna go sometime?" I said, "Heck yeah!"

After we visited there and I looked through the 12" Refracting Telescope at Saturn, it struck me like a meteorite; I wanted a telescope so bad. It was all I was thinking about. I started subscribing to popular periodical magazines on science, and of course, the Griffith Observer. I was hooked, line and sinker. All I could now fathom was, "How to become an astronaut, and when's my next launch into the stars?"

I also started to dabble in building model rockets and going to the park on weekends to commence countdowns. Now, in school, when I was asked, "What do you want to be when you grow up?" I'd say, "An astronaut or a cosmologist." I remember one prankster in class said, "I like the astronaut part, so does that mean you want to give haircuts in space, you cosmetologist?" I thought to myself, "I'm not going to dignify that with a response; for these microns, are clueless leptons." You can see that this little dreamer was now becoming a little nerd.

Part 2: FAINT FUZZIES

My birthday was fast approaching, and so was a trip. I was
planning to visit my Aunt and Uncle in the cool state of
Idaho. I so admired them for getting away from LA's city
lights and smog pollution. They had built the coolest log

cabin. I thought to myself, "Someday I'd like to do the same.
It would be super boss to bring up a telescope into that
beautiful Sun Valley, winter crisp night air."

I was so hoping for my 15th birthday, someone would get
me a telescope. "We'll see," I thought to myself. My brother
and Jim started hanging out more, and I did not mind, as I
could not get enough of the great cosmic unknown. I would
get so excited at the onslaught of each month, anticipating
next month's Griffith Observer magazine and skip right to
the section at the back, highlighting that month's celestial
events.

Unbeknownst to me, my brother and Jim were on a launch
mission to obtain my surprise gift. I acted so excited when I
received the gift; I did not want them to think I was onto
their surprise gift. I gave my brother and Jim the biggest
group hug and said, "I want this to be the best gift ever, and
you guys can borrow it anytime. If ever you want to, just let
me know. I'll wait to set it up at my Aunt and Uncle's place,
unless you guys want to look through it with me now?" They
said, "No, that's okay; you set it up in Idaho, we know that's
the best place to really see all the stars!"

Once I arrived in Idaho's Sun Valley resort area, I could not wait to set up the telescope. Most people would want to ski first thing, but not me. Then, I had come to realize, I had no clue how to set it up. I was so excited just to get it under the starry sky to experience its first light and never even opened the little wooden box it came in. I forgot to mention, a telescope's first light is a traditional moment, that is, breaking in the telescope's first exposure to the distant light photons from neighboring stars, planets, etc. It's like a christening of a ship before its maiden voyage. Now, the perplexing issue with this new 90mm refracting telescope is that it was mounted on a German Equatorial Mount, or, as I like to call it, a SOB-GEM. Allow me to elaborate, I was used to a much simpler configuration, that is, a point-and-shoot or, technically an Altitude-Azimuth or ALT-AZ mount. Simply put, up and down motion and right and left axis. This was so much easier to set up, I thought. Simple, but wait, "Help," I thought to myself. "This did not appear to be anything near easy. The GEM was like Greek to me, or at least German, and I was far from a Rosetta Stone to translate this scientific language into something meaningful."

I broke out my astronomy book and saw these lame illustrations below.

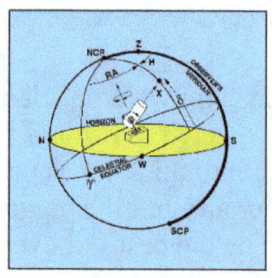 

What a puzzle it was, pointing it to the North Star – Polaris. Following the Earth's axis, wait, what. I really started to get a headache as I tried to wrap my head around this concept.

I really wanted to make it work the way it was designed; however, I felt so dumb and thought, "This is for those professional astronomers, not me. I guess I was not ready to polarize yet, so to speak. Boy, I would kill for an instruction manual or, at least someone to ask," I thought to myself again. I remember calling Jim on the phone, and he basically said, "Sorry, no manual, and I don't speak German either." We laughed, and when I hung up, I said, "Fudge" to myself. I wish they had already published the book, "Astronomy for Dummies," at that time. Oh well, I'll just set it up the easy-peasy way I was used to before. I so wanted, however, to use the cool, manual right ascension and declination control knobs (I2). Nonetheless, I got a little frustrated as I was fighting with the telescope mount, and mother nature was slapping me with some clouds. Not to mention the contortion position I had to undertake using the finder-scope. I felt like a pretzel.

Equatorial Mount with
Manual Tracking Controls

I2 – GEM With Knob Dials

I pouted and packed up my gear in its pretty little wooden box. I then decided to listen to the "Landslide" song, as this is what it felt like to me at this point. A tear rolled down my cheek. I was so hard on myself, as we all normally are, as I'm my own worst critic.

I went back home and described my dilemma to all who would listen. Jim said he'd return it, no problem, and not to worry. I felt like a failure but somewhat relieved.

The next day I saw him, and he said, "Come here, I want to show you something in the backyard." He opened the sliding glass door, and I was presented with, standing oh so beautifully, the most majestic telescope I'd ever seen. Well, of course, next to the one I saw at the Griffith Park Observatory. Before me was an Alt-Az, 4" f/15 Edmund Scientific refracting telescope, and it was so cool. It was a stunning, tall telescope on 5' oak-stained wooden tripod legs. It came with 3 Kellner eyepieces and a Barlow lens. Now, my heart jumped out of my chest with such gratitude. Jim put his arm around me and said, "I have a challenge for you to find the Andromeda Galaxy."

I knew this was our closest galactic sister to our Milky Way galaxy. It was also known as Messier 31. Charles Messier was a French astronomer in the 18th century and catalogued 100+ objects. These were identified for his colleagues, not as comets, but as celestial objects that have static coordinates. Whereas comets are dynamic and are moving through our solar system. He so loved his comets, as he was a comet hunter. These comets are visitors from the outer edge of our solar system. Once I found out he was a comet hunter, I declared that I wanted to discover one of these visitors as well, or at least see one in my lifetime. I said to myself, "Please universe, make it so…"

That very same evening, this majestic beauty of a telescope saw first light, as I explained earlier and all, and you'll never guess what I found. Yes, the first object was, of course, the Andromeda Galaxy. I used an archaic method, star-hopping. That is, in the very light-polluted skies of Van Nuys, I did see some brighter stars. This star-hopping method is where you take your star chart and locate the object. Then, identify the constellation where the object is located and find the nearest star to the object. From there, locate it in the finder scope.

Eventually, you bag your faint fuzzy object, or not. In my case, I started with the Andromeda constellation, just below the great square of Pegasus, using the above method. The Andromeda Galaxy, (M31). This galactic puzzle was now in my back pocket. This was now my favorite object of all time. You could actually see the galaxy's arms. I thought to myself, "Wrap your arms around me, you faint fuzzy, you."

Jim, my mom, grandparents got into an argument, and we never saw the talented, illustrated young man again, unfortunately. This was my first life lesson, that is, about friends and life as grown-ups see it. They decided he was too old for us, I guess, and weird that he wanted to hang with us as kids. We begged to differ.

Part 3: STAR WARS

In a Galaxy Far, Far Away...
"If I had seen further, I would be standing on the shoulders of giants..."-Sir Isaac Newton

The thing with Science-Fiction movies, there is very little basic scientific principal knowledge, if any, and lots of fiction. Other than 2001: A Space Odyssey, where it insanely represented a realistic depiction of life in space. For example, sound, like explosions in space, how do you hear them, not possible, right? There's no sound in space. Do not get me wrong, I find some of them very entertaining and sometimes there are deeper messages embedded in the narrative.

Now, I could say I totally geeked out at this juncture in my life however, I am going to get a little philosophical: Fearlessly drawn to the heavens and always questioning, "Where did we come from?" Let it flow, let it go. Keep the positive energy active in my mind. Take a soul vacation, astral project and admire the awe and beauty of the creator's vast canvass. With each brushstroke at His hand, He paints with perfection, this heavenly masterpiece. Drink it in. Be curious and keep exploring, it's what makes me happy. It's the great anticipation of what-ifs that keeps me going. As antiquity would also teach me, so many have gone before us and constructed the building blocks for us to utilize in our individual quests. So, build on them, what are you waiting for? So many twists and turns in my life. I hope to do it a service and apply some poetic license to this incredible journey. Ok, there it is. I wished upon a star and found many more...

With the advent release of Star Wars, the power of the force was tugging on me like a gravitational wave in a black hole. Begging for a new discovery within me. Or was it something else? Adolescence perhaps. I was weighing out

14

the light and dark side and of course, the attraction of girls had entered my mind. I did feel empowered by finding the Andromeda Galaxy and now I wanted to seek out, "*New life (females) and new civilizations…*", oh that's a Star Trek quote. Back on-track, so my dilemma, I had to make a choice, nerd, or charmer. How about both, and I should take an Astronomy class to figure this out. At the very least, hone my astronomy skills and knowledge and maybe impress a girl along the way.

Mr. Waterman's Astronomy class was a kick. I remember watching old movies about the Gemini and Apollo space missions in his class and was overcome with boredom because I was a teenager, and I knew all about that. From Mutnik to Sputnik, Buzz to Neil, "One small step for man, one giant leap for mankind…" I wanted to learn about the kind science, not the history of space. Finally, a break in the monotony, a classmate started whistling the theme from the TV Show, HR Puff and Stuff. He then started adding his own words:

"Just a dream from yesterday. A boy and his magic golden flute. Heard a boat from off the bay. Come and play with me, Jimmy. Come and play with me and I will take you on a trip with me on LSD…"

Maybe he was off-base, looney tunes but, I lost it and started cracking up. I got called out by Mr. Waterman. He said, "What's so funny?" I said, "I had a funny thought about eating and evacuation in space. How is this accomplished Mr. Waterman?" He said, "It's all about suction and being very careful." The whole class laughed. Then he said, "You are all invited to a star potty tonight." He was from Boston and had a thick accent. I laughed inside, as it was appropriately timed with my question, "Potty", I thought, what maroon. He was a short man and thought he was tough as nails, with his blow-dart gun he had in his class and all. I saw him more like a spaz. "So, let's see what the star party was all about." I thought to myself and said out loud, "Where and what time Mr. H²0-man?" He said, "It's right here on deck just outside of my classroom and we'll launch the sweet-16" Newtonian reflector telescope." Newtonian because Isaac Newton invented the configuration. Like Galileo invented the refractor scope and was the first to discover mountains on our moon.

I said, "Great, be there or be square." Looking back, I sure said some dumb things. Hindsight is 20/20 and please, not the year. At that moment another classmate approached me with a thick accent like the teach, and he said, "You really gonna go to the star-potty?" I said, "Yeah, I really want to look through that big scope. But I've looked through the Observatory's scope and don't think it will be much of a match, if at all. Hey, where you from?" He said, "I'm from Rhode Island and my friends call me Rabbit." I said, "What's your real name and how old are you?" He said, "Kirk and I'm 16, what's in a number dude? Hey what's your name?" I said, "Chris and I'm 15. Hey you drive?" He said, "You bet your bottom dollar, but if you're going to that star potty, you have to promise me you'll arrive FUBAR." I said, "What's that you say?" He said, "Don't worry about it, I'll tell you later. Wanna hang out after school and what observatory are you referring to? I live just across from the paak. We can meet at the Tennis Courts." I was laughing inside on the way he pronounced, park. I said, "The Griffith Observatory and sure let's meet there, after school at 3:30pm, let's synchronize our watches." He said, "Get out of here, you nerd, why I outta." We had a good laugh, and I could not wait to hang out with my new friend. This was the beginning of a long friendship, and we are still in touch, to this day.

We made up secret code words, for example, "C14" was we need to stargaze tonight and of course get drunk. C14 was the biggest commercially available Schmidt-Cassegrain Telescope (SCT). It was a type of telescope you could buy at that time, by the OEM, the stellar apex manufacturer, or the big "C". We did hang out and look through Mr. Waterman's scope and it was uneventful however, it was still fun hanging out with my newfound friend, under the stars. By the way, we did get as high as a kite by then, thanks Elton.

One lonely cloudy night I started searching in the back of the Recycler newspaper and found the oddest advertisement, it read, "Interested in forming a group of enthusiasts who love the stars...Call Celeste..." I thought to myself, "Could this be a joke or was this for Astrology? I didn't like Astrology at the time but, at the time I found sun-signs did show potential in courting and all. Well, what the heck, worth a try and can't hurt.

So, I called her and left my information on her answering machine and just like that I received a call back." I was amazed to hear, "Hello, this is Celeste, is this Chris?" I said, "Aboleelee", I was a bit tongue tied. She continued, "I wanted to invite you to my star party this weekend on Saturday if you're interested. It's also my birthday party as well. Please don't bring a gift, I just want people like you at my party." I said, "Great, what's your address, what time and can I bring a friend?" She said, "Is he an astronomy enthusiast as well?" I said, "Yeah, he loves the stars as much as I do and loves to get FUBAR'd and has a funny accent." I still did not even know what that meant. She said, "Great, here's my address and I'm over the hill. I thought to myself, "How old is she, if she's over the hill? What the heck." Then she said, "Oh and tell your friend to bring me some. I thought again, "Am I on a different planet, what is FUBAR?" Then she said, "Arrive whenever, it will be an all-night thing. By the way, what kind of music do you like?" I said' "I like Rock & Roll. Bye for now." I hung up and did not want to say any more stupid things.

Again, I thought to myself, "Wonder what kind of music she likes, and she seems interesting and young, not over the hill. Oh well, here goes nothing, I called her back and I said, "I'm

17

down for the party and coming from over the hill or, ah, see you at your house soon."

Now, the reason I'm getting into all this is, well, we all need to ask questions like, define star party and other acronyms and what do you mean by, FUBAR, exactly. This is just a small glimpse into my adolescence mania. But I did not want to sound dumb to my new female acquaintance, too late. I thought to myself, "Forget about it, score, and heck yeah! Could I have just scored a movie star or at least some sort of rich kid artist who also likes astronomy?"

Nonetheless, it was not the kind of star party we expected. It was the sex, drugs, and rock & roll kind. She did introduce me to some new music, however. Needless to say, like applesauce as a primary part of the food group, I got sauced. I stumbled around, as I drank too much and dabbled in the dark-side, and mixed up, beer, hard liquor, etc.

I found my way back miraculously to my friend's car or, the starship I liked to call it. I figured out that somehow, she was there with me, that is, Celeste. The beginning of blackout drinking, big red flag that I ignored for some time. I just wanted to break-out my telescope, so I could impress her and show her a real star party. I just remember I kept saying, "Where's my friend?" She grabbed my face and said, "Shhh I'm right here dummy." I kissed her but was really not into it and by this time the Earth was spinning off its axis.

Far, far away in another galaxy, an astronomical stirring was now in my gut. I asked myself, "Don't be feeling sick? Was this feeling called FUBAR? Well, I don't like it." It was like a Jedi's Light Saber, the way the projectile left my body. I remember thinking, "Help me Obi-Wan Kenobe, you're my only hope, or at least, where's my friend Cap'n Kirk?"

Onto the next adventure scene as this one did not go as planned, in any shape or form. We had a star party on the river, waiting for Mars, that is, the warrior planet was to rise on the east-end of the riverbank, just before sunrise. Girded up with my weapon of choice, my telescope. Most of the time we'd all fall asleep waiting for the awe-inspiring astronomical event or, you could say too many beers made me do it, blame it on the alcohol.

Another time we camped across the street from the mansion, at a neighbor's house, telescope ready, facing East once again, this time FUBAR'd, with a blender full of Margaritas helped pass the time, looking for that shaker of salt. The warrior was rising, and he was not impressed with all of us falling asleep once more. Mother nature disciplined me with, what appeared to be the longest Winter and scarce, were clear nights. Maybe I should stop drinking, not.

Sure Celeste and I hung out more and all was not lost for a time. We went to the Observatory during many full Moons. We experienced Laserium, "Dark Side of the Moon" by Pink Floyd and it became my goto music and these were the kind of moments I lived for. She explained that she saw James Dean walking around in the Planetarium and was sitting right next to her and I said, "Pleased to meet you James. Where's Marilyn Monroe?" That was that and off to the races with the next adventure. I didn't see Celeste again after that introduction to James.

My bro's spread their wings out on their space journeys. That is, Kirk joined the service, and it was just then that my grandad and I reconnected once again, like when I was a little kid. My little brother went headlong into sports. I went straight into the light side of the force looking for those distant photons and a binary-star sunset like Luke witnessed on Tatooine, that is, if I can stay awake the whole entire night, for goodness sakes.

Part 4: SPACE AND TIMEWARP

Let's do the time warp again.
Now, it's just a jump to the left…

I don't remember what happened to my telescope as I was now obscured by clouds, both physically and mentally. I needed to get my head on straight and clean up my act. This part of my life could probably break out into a whole new book, so I won't get into that now. What I do know is that my astronomical quest for star fire was now just a distant candlelit memory. I needed to leave the riffraff behind and cease participating in FUBAR. I needed some "me time on the island in the stream," maybe 40 days and nights, I contemplated. I slowly drifted off quietly into a distant Zen-land for a while. That is, I cleaned up my act, detoxed my temple, and delved a bit outside of the 3rd Dimension into metaphysics. However, I did not stay there long enough. Should have been longer, as the road to hell is paved with good intentions, as the saying goes.

After breaking through to the other side, a fork in the road appeared, so to speak, and a motorcycle accident that left me paralyzed, or at least that's what the doctors were saying. Did I acknowledge their statement of paralysis? Oh hell no. I was also now a near-death experiencer. I awoke in a tunnel-vision fog only to see my mother and stepfather at the other end of the tunnel. I reluctantly made my way back to my broken body, as it was still in shock and was not feeling pain, thank goodness. It was a crossroad and a long road to hell in my recovery. I took care of a lot of unfinished business at this island juncture of my life.

I stood tall, with crutches in hand, and started attending junior college. I thought, "Carpe Diem," and was on course with some science academics. I was now back to pursuing my interest in astronomy.

I remember my professor, Mr. Barlow. Just like the name, he was a magnified character. He would always say, "I can't wait to dismiss this class, go home and work out. Exercise my big guns with my 12-ounce curls." Then, he would gesture like drinking beer. The class laughed.

He also mentioned a perfect day in his life where he got to do more of those 12-ounce curls, with his astronomy professor. He went on to say, "There was this one day where Professor Dingbat was drawing the solar system on the chalkboard, starting with our closest star, our Sun. He drew an enormous circle with his arm in a clockwise motion. He started with the next in line, Mercury; however, he was astonished.

Now just a jump to the left, to observe his presentation of the Sun he just drew to his scale. He asked his assistant to get him the big protractor from his office. He measured his drawing of the Sun and to his amazement, he looked at the class and said, "You are all dismissed." Someone raised their hand and said, "Why, professor?" He went on to say, "I just drew a perfect circle with no tools, using just my hands, and I need to take these mitts out to celebrate. Thank you all, and goodnight, Professor Dingbat out!" Everybody laughed. I loved that class, and his sense of humor was, well, out of this world.

Fast-forward to 1995, the American dream acquired, with all the amenities. Good job, house, wife, kids, dog, and a cat. However, this one song kept playing in my head, "Once in a Lifetime," like a broken record, over and over, "...how did I get here?" I thought to myself, "I need to get centered, and wasn't it my wife who said, you liked astronomy, right, why don't you get back into it? Perhaps I should rekindle an old hobby."

I was working my rear off, that's for sure. I had a good household and helped raise some good kids. Something always felt missing, however, like a fear of missing out on something. They call it FOMO today. I read an article in a magazine at a doctor's office, it went something like this, "...This Comet could be better than Halley's Comet." I thought about it again, "Whatever happened to my old telescope?" Back to the Recycler newspaper and found an intriguing advertisement, "For Sale, Cave Telescope, may need some work on GERMAN EQUATORIAL MOUNT and Clock Drive not sure if it's working, make an offer..."

I thought, opportunity knocks, let's spark this fire again and hopefully get this telescope up and running by the time this Comet appears. What was it again, Hale-Bopp discovered it, I think. Oh yeah, and the scope was mounted on a GEM. Darn, not to worry, I'll figure it out this time. I'm a techie now for God's sake. In college, I had also picked up some computer courses.

I remember the purchase vividly. I bought it from a programmer-scientist who had a mainframe computer in his Santa Monica apartment. When he opened the door, there were stacks of computer paper, card readers, and peripherals all blocking the door. He answered the blocked door and said, "You're Chris, here for the telescope, right?" I went through the door sideways and could feel the intense heat from the mainframe. I said, "Yes, you must be Larry." He said, "Yes, and let me get the telescope for you." I expected him to say, "Excuse the mess!" but he seemed to be very preoccupied. As he was digging through his mess for the telescope out on the balcony's closet, I said, "Well, at least you don't have to worry about warming the place up during the winter, do you?" He said, "Yes, true; however, the electricity bill is killing me each month. I actually need the money from the telescope to pay a bill, as I'm now a month behind." "How much is your bill?" I spoke. He said, "Would you believe just about a thousand?" I said, "Well, that's outside my budget for this purchase." He said, "How about $500?" I said, "Let me look at it first." It appeared to be very old and dusty, full of cobwebs, but the optics looked good, and I said, "Would you take $200?" He said, "It's a vintage, collector's item; how about $300?"

I came home that night with the Cave 8" f/8 reflector telescope. It was mounted on a GEM with a working clock drive. Now, remember the internet had barely launched as the World-Wide-Web, no articles on this process or even a "how-to" set it up, and forget about the user's manual. I had to clear the webs in the tube just to see the condition of the optics. This was my first experience with owning a reflector telescope. I had to think back to my Astronomy class with Mr. Waterman, as he had a reflector. How did he set this up again and all? He had a collimating tool. "Crap," I thought to myself, "I need to get this tool." I actually found some old Edmund Scientific garb that helped pave the way.

So, I ended up giving it the full DIY treatment... The next day, I had it working; however, there was a bit of star vignetting. That is, the stars were not perfect circles, in and out of focus. It reminded me of my first telescope that my granddad bought for me, where I liked it big and out of focus. I laughed to myself. "I really need that collimating tool." I had actually set it up on top of my shed, a makeshift observatory. I remember it barely surviving an earthquake up there.

Then the visitor came, introducing Comet Hale-Bopp, arriving in all its glory into our night skies. I remember looking at it in awe with my granddad. I pointed out the two different tails and described them to him and the reason for the different colors. The blue-ion gas tail and the white, comet-dust tail. It was truly a spectacular sight and the best comet we've seen to date. My granddad and I were in awe. He asked about looking at it through my scope, and I explained to him that we needed a wide-field telescope and mine was not suitable because it had too much magnification power. We looked at it through his binoculars, and it was

23

spectacular, perfectly in the field of view. Oddly surreal, not scary at all as some had described. Yet another book of kinds.

One night, I set the scope up on my driveway, and a drunk couple approached me, and I had some fun. I was looking at the moon and explained, "You can see the lunar rover, and do you see that ink dot by it, that's the American flag. Can you see it?" They believed my power of suggestion. I laughed inside.

Nonetheless, that telescope sold, and I was now thinking about a new purchase. What should I get, especially if I'm exploring the idea of astrophotography? After some research, it came to me, like a memory from that Astronomy class, thinking back about Kirk and our code words. A C8 (8" SCT f/10). Back to the Recycler newspaper. My fingertips were ink-stained by the time I was done looking.

Above first photo of the Lagoon Nebula taken with the modified DSLR.

Part 5: VISUAL AWAKENING

To start astrophotography or not…

I remember this next telescope purchase oh so well. The guy had all the astrophotography gear and this sleek carbon fiber, black C8 telescope. I obtained all the adapters - off-axis adapter, eyepieces, dew shield, heater, etc. I had an SLR camera and was ready to take some pictures. The telescope was a goto, complete with a computerized Astromaster. I love these names, but the usage, not so much. It was reported to locate approximately 10,000 cataloged objects.

Now, probably because it was previously owned, the goto accuracy was, well, to be expected. Perhaps the nylon gears were stripped, I don't know, as I never had much luck on even remotely finding objects with the computer. That is, with the straight-through 8x50mm finder scope is how I found most of the objects. Tracking sort of worked, at least for a minute. Let me explain once I got an object in the field of view at the eyepiece, I could not keep it in that field for very long without manually making adjustments. "Man, I really need tracking to work if I want to take pictures," I thought to myself.

So back to the triangulation process - that is, using 3 points of reference in the sky to set up the goto function. Now, this all may seem like a chore and hard to understand. If you're passionate like me, no biggie, you'll figure it out.

I did know the stars and constellations somewhat, as I studied the location of my favorite showcase objects, like the Andromeda Galaxy. So, I could point to them and take a quick picture; however, I had no idea what they'd look like until the development process was complete. For the pictures and colors to come out, you needed longer exposure times.

So, my dilemma with the telescope not tracking properly and the wait time for the development process, was a challenge. Most of the time, to my dismay, I was very disappointed with the results.

The good news is the telescope's optics/mirrors were very good, and no collimation needed with this type of telescope. Also, being a closed telescope system, the mirrors don't get dusty, as long as you keep it indoors and covered. Nerd talk - also, the dew did not settle much on the corrector plate; however, when it did, I had a dew heater strap standing by to prevent this occurrence.

Now, I spent a great deal of time and money and wanted to give up. I kept thinking, I'm just an amateur astronomer, a user of this gear. How much more of a headache did the genius get when inventing the telescope apparatus and tech? This is not rocket science either; how can I be so dumb? I thought, "Perhaps I could get it serviced by the OEM and really start from scratch? Then I could really start taking some good pictures. Nah, too much extra money and effort."

Then, I decided to join an Astronomy club, The Local Group of Santa Clarita. I met with the facilitator, and she was very nice, and it was refreshing to see a woman in the hobby. She had a plethora of knowledge, coupled with stories about the "Native Americans" and the night sky. She knew about all the "Greek Mythology" encompassing each constellation as well. It was fascinating to listen to her and exude so much enthusiasm.

She said, "We will be meeting for a lecture at the Placerita Canyon Nature Center on the first Saturday, on or near the Full Moon. This month we have a special treat for both the "New Moon" field trip and the "Full Moon" lecture. We will be meeting at Mt. Pinos on Saturday at dusk. Most of us are lawn-chair astronomers, so just bring your chair if you don't have a telescope. The Perseids Meteor shower will grace our sky as well, so be ready with your cameras and don't forget your blankets, as it gets cold."

I thought how exciting and, how embarrassing to bring up my malfunctioning telescope. Maybe I'll just pretend that I don't have a telescope and just bring a chair. I also brought my granddad with me, and I was happy he was sharing in these most memorable experiences.

We started out listening to the facilitator point out the sky with her green laser, and I made a mental note - "get a green laser." She was also talking about the constellation of Cassiopeia and how she was a beautiful princess, and the warrior Perseus, as he was attempting to be her suitor. She went on about asterisms, like the "Big Dipper" and the "Coat Hanger". That is, you could look through your binoculars and find these groupings of stars shaped like a coat hanger, for example. There were many of these asterisms in the sky. I was in awe, and my granddad asked me to find these for him, and I did.

We started walking around after nightfall with our red flashlights. Red flashlights preserve your night vision. I heard this one guy talking over the crowd; he was somewhat shouting and said, "I have found the Lagoon Nebula in my telescope. Come take a look and my telescope is For Sale!" So, my granddad and I worked our way over there and fell into line.

Once I got my turn, I was speechless. I could feel this floating sensation, as if I was starting my orbital approach into the Lagoon Nebula with my spaceship, into this incredibly beautiful grouping of nebulous stars that resembled Gilligan's Island lagoon. With the dark matter lane in the nebula, it made it look like it was the water in the lagoon. I loved these very clever names given to these heavenly objects.

This was my first experience in actually seeing what looked like, a Hubble Space Telescope image and a sizable purchase for the Coulter 17.5" f/4.5 big Dobsonian telescope.

By the way, I did meet John Dobson, the very colorful Dobsonian telescope maker and I met Stephen Hawking at one of these full-Moon lectures. Exciting and most memorable moments. I'm so blessed to have met them both.

Part 6: FRIENDS INDEED

Birds of a feather…

At this point, I had not made many friends in this stellar hobby. Don't get me wrong; I adored bringing my grandad along. As the seasons changed, I did not want my grandad to weather the cold with me. I loved the crisp winter air and the longer nights. I could easily tolerate the cold as I run hot. I'm wired that way, and probably mostly by the adrenaline rush I feel at the eyepiece each time.

The Local Group had a star party outing scheduled for the weekend at Vasquez Rocks. I heard of the many movies filmed there and wanted to stargaze there as well. I thought of it being a good place for all outdoor activities as well, if not for all the dust and all. "Not good for the instruments. What the hell, I'm a giver and I need my sky fix." I thought to myself.

So, I ventured out to set up my baby, and I was proudly the largest scope there. I heard some sort of murmuring and laughing in the group, next to mine, and perhaps partaking in some extra-curricular activities, if you get my drift. I thought, "I'd like to get to know those guys." I overheard their names as Greg and Sean. One had a C11 telescope, and the other one had a C14. I heard them calling out to each other in excitement as they were finding some very obscure, faint fuzzy objects and curiously, finding those objects fast. I was enticed by their vibe of excitement and their stellar hubris. I just had to go introduce myself; however, a huge line was forming at my scope. I was already looking at the Great Nebula in Orion, so unfortunately, I could not break away and approach those guys at that moment. I thought, "Maybe I'll get a chance later."

I was using my wicked green laser to point out other objects for the folks in line. I continued to look at and describe the

Great Nebula. I kept hearing a lady saying something about an ominous object, but I refused to be interrupted at that moment. I went on describing the nebula and how stars are formed/born there, a lot like our own solar system was, and that it was relatively close to us, in the grand scheme of things. "It's only about 1,500 lightyears away," I explained. People were in awe and thanking me for the views and information. I wanted to now go and talk to those other guys.

This one lady kept saying that she was seeing a UFO and insisted on borrowing my green laser to point it out to the group. I did acknowledge her and reluctantly let her borrow it and let her know to be extra careful. Thinking I could now break away from the group for a moment to chat with those guys. Then she got so excited and started whipping the laser around on the ground, and none of us saw the UFO she was describing. I had to act fast. I told her, "You are reckless and irresponsible, and you are lucky you did not blind anyone. You lost your privilege, now hand over the laser." Looking back, I could have handled it very differently.

I then heard the excitement die down, and everyone started packing up their gear. I heard that the Local Group was now planning on going to RTMC, Riverside Telescope Makers Conference, in the Spring for their next field trip. I put that in my back pocket and thought to myself, "Mental note, register for RTMC."

 The California Poppies had now beautifully taken over the local hillsides, and Springtime was in the air. I wanted to see what-gives at the RTMC conference and again, hone my hobby skills, especially where astrophotography was concerned.

I decided to make the trek to Lake Arrowhead, where the 3-day big event was taking place. I had registered online for two nights. I was figuring I would not sleep much during the night and get some rack time during the day. I would hopefully muster up enough energy to attend some astrophotography sessions in the afternoon and a star party at night. At least that was the plan. I did not bring my pop-up tent and thought I'd rest up in the communal chalet and make some new friends, while I'm at it. Low and behold, it was a snore fest every time I entered there. So, I abandoned hope of any sleep.

Backtrack a little, I drove up on a Friday night after work and arrived somewhere around midnight to find a huge line of cars parked outside the main entrance gate to get in. Vendors and astronomy enthusiasts were all in their vehicles lined up. I was excited to get out of my truck, once in line, meet and greet some new friends and catch a glimpse of the sky. I noticed everyone in their RVs ahead of me was, well I imagined, were fast asleep. I thought to myself, "WTF, whiskey-tango-foxtrot, how can this be? It's a clear night, why is there no one out like me, under the stars?" So, I decided to walk the line of cars to the front. I was obviously too excited to sleep.

Then I heard some people laughing toward the front of the line and was eager to see who it was. Then I heard those two familiar voices again and one called out, "Oh my f'n God,

Greg, you have to check this out, it's so sick!" It was the familiar voices from the Vasquez Rock's Star Party. I was so happy to find these psyched dudes awake. At these Star Parties, I had come to realize you recognize people by their voices, as you meet up mostly in the dark. So, I called out, "Hey guys, whatcha looking at?" Sean said, "Who's asking, and we are looking through some milky, humid skies, the seeing is not good?"

Sean was no ordinary looking man; he's an avid weight trainer and resembles Arnold or, at least Jack, a lean bodybuilder, nonetheless. He approached me and said, "Wanna take a look?" I said, "Sure, and I'm Chris, by the way, and I remember hearing you at the Vasquez Rock's Star Party." I honestly was doing my best to defuse my excitement as I felt like I was going to explode with joy. Sean said, "I thought you sounded familiar. This is Greg by the way." I said, "Pleasure to meet you guys. Would you believe we are the only peeps awake?" Sean said, "Yeah, they are all a bunch of old farts and turned in early."

Greg said, "Wait a minute, Sean, I'm an old fart as well and soon to be a grandfather." I said, "Congratulations! By the way, what are we looking at?" Sean said, "Take a look, don't be shy, we are looking at the Cat's Eye Nebula." I looked through the eyepiece and, to my amazement, I could see the Cat's Eye, as it had a retina and a vein-like star in the middle of the eye. Some call this technique using averted imagination. It's really called, averted vision, where you look to the outer edges of the eyepiece. It's nature's canvass and use your imagination.

I was so impressed, and more so now that he found it without a computer or a goto mount. I asked, "How in the hell did you find the Cat's Eye without a Digital Setting Circle (Goto Computer)?" Sean replied, "Memory, kid. Stick around, you ain't seen nothing yet. I do break out the Star Charts from time to time. I have a real good one, Willie Wonka's Sky

Atlas. Want me to break out my Willie? I use these if I want to explore the super deep sky." Sean, I quickly learned, was a competitive soul and was a real kick to be around. He was meticulous with his pristine instruments and could run circles around most folks in the hobby with his star-hopping prowess. Greg, on the other hand, was very quiet and humble. "These two were the dynamic duo, Star Party animals," I thought to myself.

I said, "Wow, this is a sick view, Sean. Now, this is an 11" SCT scope, right? We really need to exchange phone numbers." Sean agreed, "For sure, Chris. And do you want to check out some galaxies now?" I replied, "Fo sho."

I don't think we slept a wink that first night. Sean said, "Dude, look for us on the field! Later, bye." Like wild stallions, dust was in the air as the gate opened, and we all raced in. Spots are available only on a first-come, first-serve basis. While I drove around aimlessly in circles looking for the chalet I had reserved, I eventually found it, parked, and power-napped in my car until mid-morning.

I woke up feeling like crap, as it was blisteringly hot, and I was slipping into a grouchy mood. However, I slapped my face and said, "Take inventory! You should now be like a kid in a candy store. Wake the freak up and get some food in your belly; that will help your mood, dude!" I changed into shorts and a tank and made my way to the field where my friends were.

The adventurous journey was getting there or rather, just beginning. There were too many distractions along the way. So many vendors and so little time. Of course, the heavy hitters were there, the unmentionable ones, and even some famous astrophotographers at the time. A husband and his wife were there as well, and what a hell of a team they made.

I quickly discovered that I was in heaven during the day as well. I thought to myself, "Will power naps suffice? Nah, sleep is overrated." I sometimes asked myself questions and answered, however chalk it up to no sleep, I'm not crazy mind you. I also heard, from people walking by, that in the evening there were going to be some presentations introducing a new line of telescopes, including their 16" SCT. I was very excited about that, but more so when one of them said, "Yeah, and the prestigious Yard Scope will be on display. I'll be the first in line for the public viewing tonight." I thought, "What the hell is a Yard Scope?"

I attended some breakout sessions and was starting to feel like a sweaty hot mess. Famished was an understatement, as excitement took me far away from my basic essential logic. "You always find yourself in this predicament, you numbskull you," I thought to myself.

Finally, I found some food and eventually caught up with my new friends. I asked, "Do you guys know anything about the Yard Scope? I heard it's here tonight." Sean said, "Yeah, we'll check it out. There'll probably be a huge line, though." I said, "Why 'Yard'?" He explained, "Because its primary mirror is 36" big. The secondary mirror is as big as, well, never mind. Also, its focal length being an f/5 is long, and it needs a really tall ladder just to look through the eyepiece." My excitement was overwhelming; I wanted to shout at the heavens. I could not wait for nightfall. Now I could get into the math, magnification, focal length formulas, etc. and all, but don't want to bore with more nerd talk, shall I continue? I digress.

That night, the usual UFO hoax was in-play at RTMC, an annual tradition, flashing blue and green lights like magic Chinese lanterns filling the sky. Or, it could have been light sticks on mylar balloons with lasers; at least that was the rumored explanation. More on that later in the Appendix.

After not too long of a wait, I was introduced to the eyepiece of the Yard Scope, and I took pause. The owner had set the scope's sight on the Veil Nebula. I was so ecstatic; I may have drooled a little. "Sorry, Charlie, I hope it did not land on your primary mirror," I thought to myself. The Yard Scope was the best beast of a scope I'd ever seen. A surreal sight for sore eyes. If you were afraid of heights, you would not have made it to the top, however. It was so, so worth the nosebleed on the way to the stairway to heaven. The Veil Nebula, I had seen pictures in books, but nothing compared to the view through the eyepiece. No wonder one part of the nebula was called the network nebula. "I get it now," I thought to myself. It was like looking at DNA brain matter through a microscope, with so many twists and

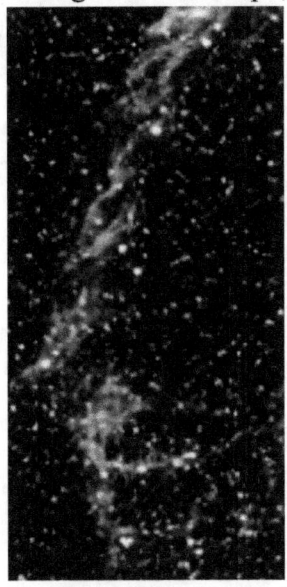

13 – Veil Nebula (Network Section of the Nebula)

turns through the neural pathways. "Holy crap," I thought to myself, "I need a bigger scope."

I continued to go to Mt. Pinos during new Moon weekends, come hell or high water. However, something felt off, and I was back to FOMO. Yes, Sean obtained another telescope, and it was different. He built it in such a way that it revealed more contrast. He accomplished this using black flocking paper and velvet lining on the truss poles. He also added a super dark, double light shroud to cover the scope, which his wife made for him. He also designed a huge thumbnail dew shield that would block any stray light. While my telescope was truly a monster in a Sono-tube configuration, Sean's used the lighter truss pole design. I was a bit envious and loved that design. Like Frank Lloyd Wright's quote, "*Where form and function are one...*" I thought to myself, "I really needed a bigger beauty and a cooler aesthetic design. Ah, where to begin?"

Part 7: OBSESSION

"Obsession is the rocket fuel for our creativity" –Jordan Ritter

Okay, now I'm off on a trip to pick up my new 25" Big Dobsonian f/5 reflector telescope in Texas. I drove over 700 miles to meet the seller at the border town near El Paso. Sandy, the seller, also gave me a custom-made eyepiece case and a beautiful 31mm epic eyepiece at the time it was the bomb or, literally the size and weight of a grenade.

I was so excited that I set up the telescope in a hotel parking lot and marveled at the views of the night sky. Even a security officer came by to check on me and he was impressed by the telescope's view of the Great Nebula in Orion. Even though at first, he thought it was a legit cannon.

I was looking forward to taking my new telescope to Mt. Pinos for a proper "first light" experience, away from the light pollution of the hotel and the "gunslingers" in Texas.

This was no longer just an Obsession telescope; it was a star-gate portal, igniting my passion for astronomy. These thoughts brought me back to Newton's quote, "If I have seen further, it is by standing upon the shoulders of giants." For me, this became my new mantra and my personal haiku. I believe that the grand architectural creator has guided me through my life thus far. "I'm determined to take this Obsession to the mountaintop and share the wonder of the universe on the canvas of space for all to see," I thought to myself, with a healthy dose of sarcasm.

On Obsession's first light, the frigid mountain conditions were nothing short of an understatement. The sub-freezing cold, with the added wind chill, created an environment that even Jack Frost might find too extreme. Yet, my determination and sheer passion pushed me through. I could care less about the cold; I was fueled by a hardcore resolve and adrenaline to achieve that elusive first light.

Mother nature, on the other hand, seemed to be having a temper tantrum. She unleashed violent, tornado-like winds. In my relentless pursuit, I braved the elements, holding onto my telescope and ladder for dear life, even as the parking lot beneath me turned into a treacherous glacier of black ice. "A headline about a man impaled with a truss pole up his butt like a human popsicle.", crossed my mind, but I was determined to steer clear of that fate. I was now literally drifting across the parking lot as if I set sail, in an iron-cross position. I must have looked like a kook, with a death wish, "But I'm not letting go of my scope and ladder." I thought to myself.

However, salvation arrived in the form of the Nordic Ranger's Search and Rescue team. Laughing off my plight, they continued on their training mission and pretty much ignored me. I said, "A little help here please." They just continued to laugh as they passed. Disassembling the scope and the ebbing of the winds marked the end of this memorable night.

With numerous telescopes, binoculars, and a growing passion for solar observations and the 2017 Total Solar Eclipse, my journey continued. I even ventured into the world of refractors, building a 5" f/15 model.

Public stargazing events became a significant part of my astronomical journey. Sharing the night sky's wonders with the public brought me immense joy, even if it meant I barely had a chance to look through my telescope. Teaching astronomy at a local high school and encountering a group interested in "Close Encounters of the Fifth Kind" research added exciting facets to my astronomical journey.

Some hints at the next part of the journey, anticipation of

Mars' closest approach to Earth in 2003 and a desire to secure a night at Mt. Wilson Observatory 60" f/20 telescope, promising more astronomical adventures on the horizon.

Part 8: MISSION TO MARS

*The echoes of your past vibrate through space, and secrets
lie beneath your broken skin; The cracks and rimless craters
on your face, Reveal to us what's hidden deep within. Now
frozen underground in super lakes, under a pitted surface
lined with breaks; A reservoir of life within our sight, to
reach for when we know the time is right. Mars, you are so
strong and so intense in your fiery red hue...*

My fascination with you, great Martian warrior, led me to
align my telescopes with your fiery form as often as I could.
Yet, my attempts often left me disappointed. No canals, no
polar caps, no surface details to speak of. I questioned
myself, "What am I missing? Is it the constant dust storms
that thwart my view?"

Even with my new Obsession scope, the view was
lackluster. It resembled a boiling egg more than the vibrant
Mars I longed to see. The quality of the view is significantly
affected by various factors, including seeing conditions,
atmosphere, and weather, particularly when observing
objects within our solar system.

It was 2003, the closest approach Mars had ever made to
Earth in my lifetime. In my pursuit of the ideal planetary
telescope, I scoured Astromart.com. However, the planetary
refractor telescopes and Japanese models I found were well
beyond my budget. I decided to explore reflector telescopes
with long focal lengths, similar to the ones I had previously
built.

I had sold my earlier telescope due to its lengthy cooling
time. The glass needed to reach ambient temperature for
optimal performance, which sometimes took hours. This
delayed my stargazing, especially when eager to observe
objects shortly after sunset. A thermal boundary layer,

caused by the heat on top of the mirror during viewing, complicated matters. I needed a new, long focal length reflector telescope to achieve my goal.

After a journey to Fresno, I acquired an optical tube assembly (OTA) with an 8" f/8 primary mirror and a curved secondary mirror cell. This configuration eliminated the star spikes and was ideal for planetary observation. In my first light test, I observed Mars, and to my delight, I could see a polar cap. This new telescope was ready for its next chapter.

I enhanced it further, building a purple carbon fiber tube, fitting a 9-point mirror cell, and repurposing the focuser housing with a dual-speed one. I added a boundary layer fan with speed control. With these improvements, I was prepared for the upcoming event at Griffith Park's satellite office, as the main observatory was undergoing renovations.

My telescope was set up on the east side of the field, closest to the rising Mars. Other amateur astronomers brought their impressive equipment and were soever present. While I had a bino-viewer adapter for a two-eyed viewing experience, it diminished brightness and contrast. Instead, I fitted it with one planetary eyepiece an orthoscopic (4-element) one with long eye-relief. I noticed a long line of people eager to peer through my scope.

As each person took their turn, they all shared the same reaction: my telescope provided the best view of Mars. Their questions came pouring in, and I explained that I had designed it for this very purpose, inspired by "My Favorite Martian." The enthusiastic crowd marveled at the view. When asked about the cost, I conveyed that in any hobby, customizing your equipment becomes a labor of love. It may test your patience, but the satisfaction of creating a unique masterpiece is unparalleled. In the words of Frank Lloyd Wright, "*Where Form and Function Are One.*"

The next expedition to Mars took us to Mt. Wilson Observatory for Part 2 of the event. The Local Group had secured a reservation for the telescope for a significant portion of the night at a reduced rate. Before Mars rose, we had a list of faint astronomical objects to explore. The night offered perfect seeing conditions, making it an astronomer's dream.

Our first celestial target was the Cat's Eye Nebula. While it had impressed me when viewed through Sean's C11 telescope at RTMC, this experience was beyond compare.

It was like receiving a visual "WOW" signal from deep space. The nebula displayed intricate, spirograph-like patterns, devoid of color but perfectly crafted in the fabric of space and time. I was left speechless and found myself holding my breath in sheer wonder.

My friend Sean, noticing my overwhelmed state, inquired if I was alright and humorously suggested that I might need another hit. My response was a somewhat delirious one, "I have never before beheld such beauty in my eye, meow." My English seemed to have temporarily escaped me.

Eventually, Mars made its grand entrance, and there was a substantial line of eager stargazers. It felt like there were at least a hundred people waiting to catch a glimpse. I turned to Sean and quipped, "I'll take that G13 hit now." We smoked out, inside the dome, filling the air with an unusual aroma and a chill vibe filled the air. What many in line didn't realize was that Mars was much brighter than usual due to its proximity to Earth, which resulted in a reddish glare of a glow that appeared magnified beyond belief. Most individuals who looked through the telescope saw what seemed like a big, washed-out red glare of an object, and their reactions held hope for us, getting to the eyepiece a bit faster.

Sean and I had come prepared with polarized sunglasses, but Sean had taken an extra step by bringing a polarized eyepiece filter. Some curious onlookers attempted to use their sunglasses, but their comments remained skeptical. The good news was that the line was moving at a steady pace. When we finally reached the eyepiece, we used the polarized filter, which turned out to be the Holy Grail of the quest. Our mission was accomplished.

What I saw through the eyepiece was, hands-down, the most incredible view of Mars in my life. The image closely resembled the Hubble picture you see below, to scale. I felt transported to another dimension and couldn't help but shout in the dome, "Holy Crap, I love you, Mars!"

As the word "Crap" reverberated throughout the dome, it brought back memories of my childhood stargazing experiences with my granddad. I recalled my first view of the star Sirius and how I had felt like an excited kid all over again. I couldn't wait to share this experience with him, but I realized he might not fully grasp it now. Dementia had taken hold of him, and his memory was fading. Nonetheless, I promised myself that I would mention it to him during my next visit.

The crowd had dwindled, likely due to the lateness of the hour, but I remained undeterred. I glanced around and realized it was just the telescope operator and me. He offered, "Wanna take a look at the Moon? It will ruin your night vision, though." I eagerly replied, "Please, ruin me!" He handed me the joystick to control the telescope, with a friendly warning to be careful on the platform. I assured him, "Not to worry, I'll be extra careful, kind sir."

I couldn't contain my excitement as I felt like the Genie in the bottle had just granted my first wish. The sky was no longer my limit; I now had the power of the cosmos at my fingertips. I thought to myself, "Oh, what incredible power I now have in my hands. My new nickname: He-Man or, he's the man."

In retrospect, it was one of the most enjoyable moments of my life. I treasured the experience as I navigated over the Moon's surface, hovering near the terminator, and then making my way to the Tyco crater. I felt like a lunar explorer, almost a "hairy lunatic" howling at the Moon, living out a role in my mind's twilight. As the Sun began to rise, I rushed to explore each crater, even venturing inside

Kepler to get a 3D perspective. Then, I headed over to the Sea of Tranquility, wondering if I could spot any other interesting lunar features, like the Moon maiden. And during this thrilling experience, another thought crossed my mind: "What would it take to become a Mt. Wilson curator?"

Part 9: BIG GUNS

"If at first you don't succeed, try try again." -Edward Hickson

So, the Obsession scope had served me well, but I noticed an issue with its accuracy. After seeking help from the astronomy community, I decided to sell it. The intense obsession with the hobby was growing, and I contemplated buying another telescope.

Some friends recommended a telescope maker who was renowned for crafting great mirrors and had ventured into creating fully automated, robotic systems. I expressed my interest in a 28" telescope model if he ever reached that level, although I had some doubts.

Then, as fate would have it, he decided to sell his personal 28" telescope to me, Spica Eyes. I didn't hesitate to make a deal and became the proud owner of this magnificent instrument. Finally, I was content with my equipment.

The first night I used it, mother nature would put the kabash on me again with an extreme cold, but the seeing conditions were magical in Joshua Tree National Park. My friends, Sean and I embarked on a cosmic adventure, exploring galaxies like Black Eye, Sunflower, Whirlpool, Pinwheel, and the Whale. We took in the night sky with awe, like cosmic sorcerers.

As the clock struck midnight, we switched to a live, color camera to observe the Great Nebula in Orion. The sight was glorious, and the colors were so vibrant that we were left in awe, despite the ruined night vision.

Ultimately, I realized that an instrument of this caliber required an unyielding commitment, and I eventually decided to downsize. I traded the scope for a C11 SCT/CDC to explore astrophotography and the next phase of my astronomical journey.

Part 10: CAPTURED PHOTONS

"A picture is worth a thousand words." - Henrik Ibsen
Life of Vincent Van Gogh's, Starry Night and a quote,
"Art washes away from the soul the (star) dust of everyday life. - Unknown"

I desire to own the heavenly art. Time to see the wizard and off into the next phase of the hobby. Back to Astromart in search of the next scope. No more health or back issues hopefully. I put my back into full swing, and I was 100% committed to capturing the heavens with my DSLR camera. Like any hobby it's supposed to be fun, not a health challenge, sheesh.

Now, the C11's goto function was absolutely spot on and was a total pleasure and a cosmic delight. That was exactly what I needed to start on my astrophotography run. So, I started with my favorite solar system object, Saturn. From my backyard, I retrofitted my DSLR camera, punched in Saturn and I took my first picture, see picture below. The C11 had a long focal length so the image scale was not too bad and zoomed in slightly with a quick shot and the first photo came into view. "Wow, not bad", I said to myself, "However, it still looks better through the eyepiece." I called myself out and said, "Now I can I make this look better, I think so chief"

Maybe using a computer application, however, I like raw untouched pictures. Backyard astronomy is "Ok" but, I either needed a light-pollution filter or, a better location. I set up the rig at Mt. Pinos and it was so much quicker than the big Dob setups. I purchased a Bahintov Mask to achieve better focus. I was now ready to triangulate the goto computer. No need to collimate the mirrors with a SCT telescope configuration. The goto computer had a GPS and the technology was so sweet. I found one star that I triangulated on before. It was actually now so simple for me.

"Chalk it up to experience.", I said to myself. I used a laptop and software and other than that, I just had to weather the cold. The rest is history and patience are all it took to get me to this comfort level. I finally reached my plateau. It took some time to produce decent photos. The standing outside was a bit cumbersome and watching the camera's tiny screen was a pain in the neck.

I then hooked up the AUX/serial port from the goto computer to my other laptop and used that one for the camera. Now that's what I'm talking about. It was legit and coming together.

I was now in my, not too shabby, toy-hauler trailer snapping away and taking lots of beautiful photos. However, I could not close the door to my trailer all the way, it was those damn cables. "I need to build a wireless, RV observatory", I thought to myself. Back to Astromart and the forums. I could not find any information on that concept either, that is, wirelessly, remote control the telescope and the DSLR. "Was it even possible?" I thought to myself. I know there was some Bluetooth technology that had some potential however, there was some new cutting-edge tech with WiDi. I thought I should research this more.

I then added two new items to my tool-bag, a wireless camera control and a WiDi for wireless telescope control. Next new moon to see if my research paid off. Well, it sort-of worked but the software was not a perfect fit. Back to the drawing board. I tried out a bunch more apps. Then, as patience is my virtue, the universe answered, and some new perfect software fell right into my lap. The telescope can now be controlled easily by a smartphone.

But wait, it was for iOS only however, I actually was still using a berry old cell phone up until this point. Then, I crossed over to the dark side, and decided to purchase the big brother. I then found this was a walk in the park, that is, easy setup, easy triangulation, easy goto, and easy photos. All from the toasty comfort of my now mobile observatory. Now this was "HiTech, Geek Astronomy - 101".

The photo above was my first capture of the Swan Nebula. It seemed like the way to go. Don't get me wrong, there were still some minor glitches along the way, battery charges, triangulation is always a bear in the cold as the

instrumentation sometimes freezes up. Other than that, cake!

I ate up the sky with my carbon fiber (Blackie) C11 and shot the Ring Nebula above. The Great Nebula in Orion above.

The Lagoon Nebula below. This is the one that captured me.

The Andromeda Galaxy, taken with the C11, so lovely.

Comet ISON, next shot before its kamikaze into the Sun, below.

There was still some field rotation. That is, some star vignette or, kidney bean effect on the stars. I did attempt to use an equatorial wedge that's supposed to correct for that, and it was not 100% effective.

I like to say, "I take my coffee, black as midnight without a moon. I'm always in the mood for a Black Russian especially while I'm star partying. Helps with the cold and keeps you awake.", I digress.

A wide-field shot, just mount the camera with a telephoto lens on top of the telescope to get piggy-back results. Onward towards the core of our Milky Way galaxy, so beautiful.

Some springtime galaxies, below. The last one in the mix is another wide-field, piggy-back shot.

Turbulent solar winds, setting sail again like with the Obsession scope, set your eyes on the Sun, with appropriate filters, of course.

Below, wide field shooting the Horsehead Nebula and the Flame Nebula.

Part 11: SOLAR TOTALITY

A "Total Solar Eclipse" parallels a spiritual experience...
"As the Sun wraps his arms around the moon,
She craves his touch like it'll be gone too soon.
The warmth he gave was burning her skin
Caressing her very soul within
As they eclipse and darken the bright sky
She can't help but wonder why
After all this time, she's been waiting for him" – Euphoria

Okay, the great USA Solar Eclipse was to arrive in August 2017. With lots of planning, we successfully pulled it off, thanks to my wife and her incredible ability to make all things possible. We made the trip worthwhile, as it was a multifaceted journey with some parallel trips planned alongside. That is, we also planned to visit Yellowstone National Park and the Grand Tetons.

I'll do my best to give it the poetic justice it deserves. There is absolutely nothing better than being there.

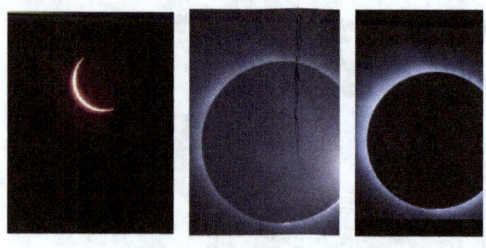

First, we made our reservations almost a year in advance. We also purchased shirts and eclipse-certified glasses. Thinking back, it really makes you want to join the club and be an eclipse chaser. Next, we made sure we were 100% committed. Our eclipse reservations had to be inside the path of totality. We made a few just in case one had fallen through, as some did. So, we can maximize all our efforts for the full duration of the eclipse. We secured our spot and had totality for just over two minutes. However, I assure you, it was the fastest two minutes of our lives. As you approach totality, the temperature slowly drops as much as 15°F degrees. During the exemplary summer noontime moment, it got surreal, and I felt a stillness I can't seem to explain. It was as if you were in a dream and floating up to the celestial event in the sky. It was suddenly night, stars and planets started to appear in the sky; however, the Sun and Moon were taking all the focus. It was visually spectacular to say the least. At one point, I removed my solar filter off my solar scope and almost passed out, as I forgot

how to breathe, once again. I was bathing in the Corona, as it was blasting out in every direction and seemed alive. NASA's shot above; it is the closest depiction of how it looked visually through my scope.

I eventually did it again, that is, sold the rig and "why" do you ask? Too heavy, and a fork mounted C11 has the tendency to present field rotation. So, I reluctantly sold the C11 and fabricated a 10" split-tube f/6 on a beautiful GEM, see below. Outside of all the guess work and geek talk, this was the new right setup and a honor to own. Of course, I tweaked it to my liking replacing everything but the OEM OTA. See picture below.

Now, just a bit of a photo gallery below:
My pictures, geek talk, range from 15 – 55 second
exposures, using only the bulb setting on my DSLR camera.

I did modify the camera to a full-spectrum system, allowing
all the possible bandwidths of photons to pass through to the
CMOS sensor.

A true time-travel experience, so to speak. The photo below
is of the Horsehead Nebula.

Above, the galaxy looks pinned to the vastness of space, hundreds of thousands of light years across.
The above galaxy, shot with my 80mm.

Orion Nebula shot with the C11.

The 10" reflector proudly shot the below Dumbbell Nebula. That star went Nova, and that's the gas remnants remaining. The central star is now just a white dwarf. Similar to what our Sun might experience after it gets past its mid-life and starts its journey to Nova.

APPENDIX

Like Einstein's ($E= MC^2$), his greatest blunder (The Cosmological Constant) on the discovery of Dark Matter. His greatest so-called mistake turns out to be one of his greatest discoveries. There's research going on every minute of the day, searching for NEO (Near Earth Objects). We even recently witnessed an interstellar visitor (Oumuamua), a cylinder oblong shape, and it perplexed what science knows about objects from other solar systems or, an extra-solar object. Its trajectory changed on several occasions, unlike our known visitors, such as comets that are on a linear path, revolving around our Sun originating from the outermost part of our solar system or the Oort cloud. Oumuamua was not. It just tumbled past our solar system, barely detected and had little impact on the gravity of our Sun. What about Extraterrestrials? What do they see that is so interesting in humans? Are they anthropologists that planted the human seed, architecting an evolution many millennia ago? If so, they appear to be like ghosts, not wanting to interfere, like Star Trek's prime directive?

Above, I caught a couple of anomalies while shooting the M104 galaxy on wide field.   Close encounters that happened to me, you might ask? Yes, there are some, and I'll do my best to describe and provide the due diligence they deserve. Some have logical explanations, and others do not. Naturally, I did a lot of stargazing as an amateur astronomer for the better part of my life, so statistically, it's more likely to happen as the more you look up, the more you increase your odds. It's like winning the lottery; you have to play first.

MY EXTRAORDINARY EXPERIENCES **8 years old, North Hollywood, CA: Playing in the front yard with my little brother and our two friends at the time, we looked up and saw a huge saucer-like, shiny metallic object hovering over us, about 25' in diameter and about 100' above. The time of day was just about lunchtime. We felt a low-frequency noise as it reverberated inside our bodies, not through the sensory perception of our ears. We were awestruck and catatonic. We stopped playing for some unknown span of time while we continued to look up. We then proceeded to play as if nothing at all had happened to us, and we never spoke of this daytime incident again. It was as if our playtime was not interrupted, nor did we acknowledge the time loss. It was as if this moment we witnessed was erased from our minds. Fast-forward 10 years later, my brother and I, as teenagers, had the same dream of this incident; this exact incident was reenacted in our dream. We compared notes and 100% believed that UFOs are real. We actually both felt we may have been abducted but don't have any recollection of being onboarded.**

Young adult, Big Bear, CA: Was driving with a friend, heading out for a winter snow-skiing trip, and was telling the story above as we drove. The story was met with skepticism, to say the least. Further along the drive, on the way to the ski resort, my gaze was turned to a white glowing, cigar-shaped object, just over the local mountain ridgeline as we started our ascent to the higher elevation. I immediately pulled the car over to get a better view. We got out of the car, and upon doing so, we witnessed an armada of smaller craft converging with the cigar-shaped, main object. They were all moving very fast around the cigar-shaped object. Some were maneuvering at right angles. Once they all landed on the mothership, as it were, they seemingly disabled their lights, one-by-one. This went on for about 5 minutes until the last white craft landed, and then the mothership just dematerialized. I then asked my friend, "Do you believe in UFOs now?" and the answer was, "Oh Hell Yeah!"

San Fernando Valley, CA: While at the NIKE Missile Base, I parked and was admiring the overlook of the San Fernando Valley. I originally was parked there with several other vehicles at the vista point. I noticed that time had passed rather quickly and that we were now alone. I noticed that there was a red glow on the dashboard of my truck. I thought at first it was a fire or an electrical short. But then, I felt a low-pitched frequency, kind of like I had as a child in my dream but coupled with a subwoofer. I turned around and to my surprise, we saw a red glowing orb-like object, and we felt very scared. It was practically on top of my truck, about 20' away at a 45°-degree angle. I think it was about the size of a volleyball. Stunned for an undetermined amount of time. Once I came to my senses, I turned back around and attempted to start the truck. Just like in the movie "Close Encounters of the Third Kind," my truck would not start. I felt the orb slowly getting closer. I said a quick prayer, hit the steering wheel with a thud, and then the car started. I hightailed it out of there as quickly as I could. It followed us for a spell, then it went in another direction.

68

Mt. Pinos, CA (Part I): My friend was with me stargazing. He had just purchased his new reflector telescope, first light at Mt. Pinos. He said, "Man, Venus is so beautiful and awesome; I never thought I'd see such details. I really love this new telescope." Then he started shouting at me and said, "Look, come quick, Venus just blew up. Oh no!" He exclaimed. I then looked through the eyepiece and saw what looked like dozens of shimmering crystal-like objects all floating around this big, central white orb. It was a surreal moment and was absolutely the best visual capture of a group of UFOs and its smaller ships that I have ever seen, up close and personal, with a telescope.

RTMC Lake Arrowhead, CA: It took place in the late 90s, a small group of amateur astronomers and I were stargazing at nightfall, and that all-familiar now, low-pitched frequency came over me again. It was an eerie still moment, once again. We all looked up at the Zenith Meridian and noticed two objects appeared to be dog fighting. One blue and one green in blurred color, as their velocity was warp-like fast. Moving at this high velocity from one part of the sky to the next and at right angles. This went on for about 15 minutes until they took on the persona of background stars and vanished. We were all so surprised at witnessing this that we could not take our eyes off of it during the incident. One of my friends attempted to look at it through his binoculars during the incident; however, he indicated they were moving too fast and could not get a steady fix on either of them.

Mt. Pinos, CA (Part II): We were stargazing the weekend before that dreadful day in early September 2001, with a small group of my friends when suddenly that all-familiar feeling came over me again. My friend Sean pointed out a craft hovering over the mountain range and just above the tree line. This cigar-shaped craft was dark and had no lights. It was about 100' long and approximately 100' away from us. We all had cameras and high-powered optics; however, we were catatonic. We all just stared at it as it was hovering there, slowly moving. Someone shined a laser on it, and then it cloaked. As it continued on its slow trajectory, you could barely make out the background stars behind it as they blurred. They looked distorted as if there was a sort of lensing effect, and then it just vanished into the night sky. The buzz was still in the air for some time after.

Some may say conspiracy however, we beg to differ. The recipe during cloudy nights, was to share some wine with friends, add music, lectures, software and of course a pinch of some sort of astronomy. Sean just downloaded Google Moon release v1.0. He said, "Dude, have you seen the structures on the Southern hemisphere of the Moon. Come on by and check it out. It will surely blow your mind!" There was structure and most was blacked-out or redacted. The next update for Google Moon smoothed out the surface and seemingly removed all the structure.

CE5-KLU (Keep Looking Up) Field Research: Close Encounters of the Fifth Kind (CE-5) Protocols of ET invitation. Download the app, network and voila. To date, I've been fortunate enough to lead a handful of these extraordinary events with a team of great people. The field research team is made up of like-minded, spiritual people. As we meditate, we remain grounded in this good Earth. We center ourselves as part of a much bigger purpose as we begin our heartwarming ET invitation with clear intent. We start by leaving behind our egos, fears, politics, and religion behind, then we open up our peaceful, positive hearts to the great beyond. The most spectacular of these events for me are the ETVs (Extraterrestrial Vehicles) that continue to flash in a particular part of the sky typically. Also known as flashbulb objects, these objects seem to enter our dimension and then phase-shift out of our dimension as quickly as they entered, with several bright strobe-light flashes. These do not resemble any of our known aircraft. We also cross-reference all these events with real-time apps for verification purposes. When these extraordinary events occur, you can feel the electricity in the air. It's a surreal moment that transcends space and time. It makes me feel so grateful to be part of this bigger purpose and awake, knowing there's more out there than just us. The hours fly by as if they were seconds during our fieldwork. We have gathered a good degree of evidence with our gear and will continue to practice CE-5.

In closing, I'll end with a Karma story. Two of my telescopes were removed from me in an unfortunate circumstance. I prayed about it and calmly asked God and the universe to return them to me, if it be His will and His time for me to receive them again. They came back to me two-fold, way better than before. This coincides with my first-hand experience. I remember hearing this quote some time ago, "Coincidence is God's way of being anonymous." I have experienced this many times in my life, and I am so grateful for His grace. All you have to do is ask, and your answer will come. The grand architect is standing calmly by, waiting for you to open the door and let Him into your life. In return, all that is required is your patience and gratitude. I hope you have enjoyed reading my story as much as I have in telling it. May you wish upon the stars, wherever you are, and see more of your stories unfold...

Dear readers, and my friends who share in the love of the great vastness of space and the great unknown:

I'm acknowledging some specific highlights in my book, that is, my passion is the pursuit of Astronomy. Exploring the cosmos is a journey best shared with those you love. I thank my friends and family that have stuck by me during those celestial pursuits and have developed a shared fascination with the vast solitude of space and now have an insatiable curiosity for the discovery of truth. Life's lessons sometimes go unnoticed or viewed as annoying. I implore you to embrace them, that is if you experience them in pursuit of your passion. Let the passion drive you and go for it. Stick to it and have fun with it and the journey, even during the most difficult times.

ACKNOWLEDGEMENTS

Astromart.com & Cloudynights.com

Authors and Novelists Everywhere

Dr. Steven M. Greer

Frank Lloyd Wright

Griffith Park Observatory & Mt. Wilson Observatory

NASA – HST & JWST

Museums Everywhere

National Park Service

SETI

Telescope Makers Everywhere

Other unmentionables in the arts, music, film and television

ABOUT THE AUTHOR

The author, an IT Senior Project Manager hailing from the
sunlit shores of Southern California's Gold Coast, finds
inspiration nestled in the Zen of the San Fernando Valley
and the sacred embrace of Mt. Pinos. Seamlessly integrating
the precision of professional expertise with the vast curiosity
as an avid amateur astronomer, he navigates a dual path of
corporate mastery and celestial exploration. With a profound
appreciation for the arts in all forms, his open mind and
commitment to truth-seeking illuminate both professional
and personal pursuits. Each day unfolds as a canvas for
inspiration, a journey brimming with eagerly anticipated
lessons and warmly welcomed experiences.